QUANTUM PHYSICS FOR BEGINNERS

DISCOVER THE SCIENCE
OF QUANTUM MECHANICS AND
LEARN THE BASIC CONCEPTS
FROM INTERFERENCE TO
ENTANGLEMENT BY ANALYZING
THE MOST FAMOUS
QUANTUM EXPERIMENTS

DANIEL GOLDING

TABLE OF CONTENTS

>> INTRODUCTION >>

We're about to tackle Quantum Mechanics (QM from now on) and it could be frightening. Unlike other arguments, however, the real difficulty is not in understanding, but in accepting something completely senseless, precisely in the right meaning of the term: not sensible, that is, contrary to the perception of our senses. It is then necessary to brainwash ourselves first and prepare us for seemingly unacceptable phenomena..

THE BEGINNING WILL, THEREFORE, BE DEDICATED TO THIS RESETTING PHASE. As Gilmore said, we're entering Alice's wonderland: the quantum world. However, you will notice that quantum mechanics is much "easier" than the theory of relativity. In fact, you could get a child to help you digest certain concepts. The great difficulty does not lie in their complexity, but their absurdity in terms of logic acquired after many years of existence in a world that constantly follows certain rules. The more the brain is free of preconceptions and ingrained notions, the better it is.

» CHAPTER - 1 »

RESETTING OUR BRAINS

I n ancient Greece, there was a very heated dispute about the nature of light. Claudius Ptolemy and other scholars thought it was a kind of ray traveling from the eye to the object being observed.

It seems, of course, a huge misinterpretation and this was proved only a few centuries later. However, with the advent of QM, this conclusion has become much less absurd and naive, because of the subjective repercussions of the phenomena concerning the infinitely small and, therefore, the light itself.

I used this somewhat ambiguous and mysterious beginning to tackle the fundamentals of the QM precisely to get as close as possible to the amazement and sense of uncertainty and incapacity that a human mind, even of the highest level, undergoes when it approaches the rules/non-rule that govern elementary particles and their absurd world, truly similar to "Alice's Wonderland."

The important thing is then to be able not to be surprised at anything and to accept and try to describe even what our daily logic would tend to reject. We must not, however, pretend to understand what is happening in the QM. It is by definition incomprehensible, as the 1965 Nobel Prize in physics, Richard Feynman himself said, precisely the one who can be considered the greatest illustrator and divulger of the QM problem in its most essential and fundamental points.

His double-slit experiment is considered by many to be the most beautiful and elegant physics experiment ever made. Yet, the conclusion is the least scientific one can think of if one remains anchored to the physics that regulates the macroscopic reality of the Universe.

To introduce the QM and make it less absurd than it might seem when compared to everyday macroscopic reality, it is convenient to start from light and its history. A story that seems less than scientific. A continuous change of ideas and irrefutable evidence colliding with each other, apparently annulling the previous conclusions.

It was considered for a long time as a shower of particles or corpuscles, similar in some way to the bullets fired from a rifle (Newton himself was more than convinced of this). After centuries of long research and experiments, it became clear that the situation was not to be that way: the light behaved, beyond any reasonable doubt, like the waves of the sea. The phenomenon of light interference and Young's experiment, in particular, provided incontrovertible proof.

In the twentieth century, after other long researches, light appeared, however, again composed of particles, called photons and, through the photoelectric effect, it was even possible to count them (the Nobel Prize taken by Einstein was not given to him for his theory of relativity, but precisely for his studies on the photoelectric effect). The game, however, was not over. Other research quickly demonstrated that the undulatory nature could not be completely erased. Even if we are talking about ideas and deductions made by the greatest scientists of humanity, what was achieved was a misunderstanding. A misunderstanding so obvious and evident that it can be noticed by anyone.

Allow me, then, to accompany you towards a thought that I consider very important to make simpler an argument that appears not so much complicated as really absurd. We are used to living in a world where everything we perceive is composed of macroscopic events and complex systems.

Classical physics has always tried to describe and explain these phenomena, succeeding very well. Laws have been written and experiments have been made to prove them. Everything seemed almost perfect. Not everything was (and is) perfectly explained — but if only — it would be enough to continue in that direction, and sooner or later, the laws of physics could describe the whole Universe. On the other hand, what is still not understood is essentially linked to phenomena and increasingly gigantic objects. It would be enough to continue with perseverance according to well-established logic and experience. And, instead, things are not like that.

You have to look through the telescope from the apparently wrong side to see it. In other words, to explain the bigger things you have to look at the smaller things. Not only that, though. A microscope is not enough. You also have to accept situations that appear completely senseless in the macroscopic reality perceived every day. And this is undoubtedly the simplest, and at the same time, the most difficult part of the whole enterprise.

Time as well assumes an ambiguous and incredible characteristic. We know very well how to describe the difference between the past and future. We know from direct experience the arrow of time. Yet, the laws that regulate the essence of the Universe are all reversible, that is, they have no problem in replacing past with future and vice versa. How then is it possible that macroscopic life follows an irreversible verse? Isn't it, after all, composed of microstructures, governed by reversible laws?

The real key point of the whole QM is, therefore, the confusion. Throughout its history, man has transmitted, through phenomena perceivable with the senses, information to the brain that could be explained by the creation of theories that, in turn, could be verified directly with the senses. In other words, a man followed the Galilean scientific method. Trying to go towards the infinitely small, instead, you have come up against something completely new. The senses witnessed absurd phenomena and transmitted to the brain information that lacked logic and common sense.

Poor brain!

It's gone into a deep crisis. How could he explain something impossible to verify? The only possibility was to imagine a completely unexpected reality, without the help of the senses. He had to build theories based only on mathematics, without being able to see the actors of the phenomena directly. Light was precisely one of the most "visible" and irrational absurdities.

Today we know how light, its photons, and electrons, fundamental pawns of the Universe, behave. However, it is very difficult to say what they are. If I say they are particles I might give the wrong impression; the same thing would happen, though, if I said they are waves. The exact way to answer is to say that they are something that can only be described through the QM, the only one able to break down the barriers that our daily experience obliges us to build. Photons and electrons behave in a way that no one has ever been able to see. In fact, we already know how to use it for everyday purposes, but we absolutely cannot perceive its nature without a purely mathematical vision.

To keep it simple, we will, therefore, have to confine ourselves to the basic concepts. Any attempt at an in-depth description needs purely intellectual mathematics. That would be the product of our brain and not the repetitive experience of normal phenomena.

The atom can continue to be compared to a miniature planetary system, knowing, however, very well that it is only an artificial and partial representation of an unrepresentable reality. Unfortunately, one cannot even try to go around the obstacle and avoid the bizarre behavior of photons and electrons: all the particles that form the Universe follow the same rules.

If we want to fully understand any physical law that surrounds us, we have to go through this absurd world. Like saying that to get out of the house and into reality, we would always have to cross "Alice's Wonderland." We could do without it, as we have done for centuries and centuries, but once we know that what we think we understand with our senses is only a partial vision of everything, how can we ignore it?

Also, keep in mind that the difficulties of the QM are essentially psychological. We can almost completely exclude the formulas but the concepts remain absurd. The first obvious answer would be: "Is it possible that this is the case?." The brain immediately tends to look for an explanation related to everyday experiences.

And this attempt must be fought and eliminated immediately. It would be useless and harmful to try to describe the "things" of the QM through a series of events and connections related to our sensory range. It would be impossible. One can only describe them as they are, trying to free oneself from all remorse and clichés.

Let's imagine "resetting" our brain and immersing ourselves in a world that we cannot yet know and that we may never know, because of its very characteristics.

Let's not create alibis, though. When the theory

of relativity appeared, it was said that only a very limited number of people understood it. Actually, that wasn't the case. The very limited number of people who understood it was simply made up of people that had read something about the theory. When it started, it suddenly became a subject within reach of high school students, and even less so.

The same is true for QM, although the intellectual effort is not helped by equally explanatory visual representations. And it is a pity that in schools there is still no attempt to "breed" brains to reason on an apparently unreal, but concrete reality.

Let us remember once again that QM and its implications and predictions are now normally used in everyday technology.

Reminds me a little bit of the history that was studied in my day. It always stopped at World War I. We didn't even want to talk officially about what happened afterward: it was better to ignore it.

Many professors' poor knowledge of QM makes it preferable to exclude it from physics courses. Yet, it is not something purely theoretical, all to be confirmed, but an unreal reality so true that its laws and predictions are used daily. Sounds like nonsense.

As already mentioned at the beginning, the great Feynman, essential reference for the description of QM, said: "Nobody understands quantum mechanics." And that is the sacred truth. No one can understand it precisely because no one knows anything that resembles it.

The same scientists who describe it illustrate, after all, a personal vision of a reality unattainable by definition by our senses. In other words, everyone tries to express concepts, now established, according to their imagination. The strength of mathematics is just that: it allows us to unify infinite subjective visions in a form intelligible to all.

So, dear boys, we're in good company. Not understanding MQ is already a good place to start. Being able to describe it will be the only real effort of our brain, clean of the notions stored so far.

Forgive the repetitive sentences and redundant concepts, but it is essential to get off on the right foot both to understand the descriptions and to fully enjoy the sense of mystery and beauty of the microscopic world.

Feynman used a descriptive method that is exciting and simple and remains perfectly concrete. It is based on analogies and contrasts with everyday reality. I will try to follow his approach. First, however, it is good to remember what is the phenomenon of wave interference and Young's solving experiment.

If you understand these two notions, everything else will flow into... the most unexpected absurdity. Don't worry, though, about not being able to understand the conclusion of Feynman's experiment, since even he said that it is impossible to understand it. We may, therefore, feel that we too are great scientists since we also can easily understand very little.

As already mentioned, this result is already a fundamental step for a more accurate description of the QM. If we said we understood, using the notions of classical physics, we would surely have taken the wrong path.

You will notice, in a moment, that I will describe the phenomenon of interference twice. A pointless repetition? Perhaps yes, but sometimes reading the same treatment twice, expressed in slightly different words, can help a lot in understanding. Besides, I didn't want to modify Feynman's experiment too much, even though I described the interference beforehand. As we shall soon see, one plus one is not necessarily always equal two.

CHAPTER - 2

WAVE INTERFERENCE

Many practical examples perfectly describe the phenomenon of wave interference. Take the sea waves, for example. Everyone knows that if there are two sources of waves (two stones pulled into the water) when they meet, they can add up or cancel each other out. The waves interfere with each other. It is said that interference is constructive when the "crests" of the waves meet, giving rise to larger wave amplitude, while it is said to be destructive when a "crest" and a "trough" overlap. The famous anomalous waves are precisely given by sporadic and occasional encounters of waves of normal amplitude.

Let's summarize this well-known phenomenon through the following image:

If we throw two stones into the water, two waves come together and add up and cancel each other out. In the image below, we see in detail how two waves of equal amplitude, which add up while they are both at maximum, give rise to a wave of double amplitude. The interference between them is constructive.

Obviously, if the waves meet "out of phase," i.e., when the maximum of one corresponds to the minimum of the other, the resulting wave nullifies, i.e., the amplitude is zero. The interference is destructive. Even if the two wave systems have different amplitudes, we continue to sum them up, considering negative the amplitude of a "trough," i.e., the resulting wave has minimum amplitude but not equal to zero. From now on, however, we will always consider waves of equal length and amplitude.

To make it clear: when in phase, the two waves create constructive interference, resulting in a wave of greater amplitude. When 180° out of phase, they create destructive interference.

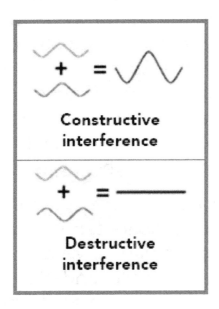

Constructive interference

Destructive interference

In Young's time, this phenomenon was already known, although in a very empirical and not generalized way. Surely waves were known, but they were associated with macroscopic phenomena such as the sound. Somehow waves were a "something" able to get around obstacles. If a wall was placed in front of a rough sea, it was obvious that the water could go around it and spread to the other side. In the same way, the sound and source of the noise can be heard even if there were a mountain in between us.

Surely the light couldn't do that since it only spread in a straight line. The decisive proof was in the sky: if a planet passed in front of a star, the star's light would disappear and it would certainly not be able to circumvent the obstacle. Moreover, while in the sea waves there was the sea and in the sound waves, there was air, two substances that were used to propagate the waves, in the cosmic vacuum there was absolutely nothing. Those who tried to think about the waves of light necessarily clashed with the concept of "ether," that is something that could make the light waves propagate.

Young knew very well that the task of proving that even light propagated by waves was not easy and it took a truly unassailable experiment. Even if it went a little undertone, especially because of its shy character, Young realized a real masterpiece: the double-slit experiment, which will then be resumed and modified by Feynman himself.

A beam of light rays (we keep calling them like this, for now, since they are good for both particles and waves) hits a screen where there are two very small holes or cracks, which can be considered the origin of two new beams.

The two holes or slits become, in fact, two independent but perfectly homogeneous light sources, since they are created from a single homogeneous beam. At this point, we put a screen that collects the light coming from the two holes. With certain amazement (for the time it was made), you can see light and dark fringes, very similar to the waves of the sea coming from two different sources. The only difference is that, in the case of water, the observer looks at the interference figures from above, seeing two series of circles that overlap or cancel each other out, while in the experiment with light he has to observe the interference figure that forms on a screen.

The experiment is schematically shown in the image below (seen from above).

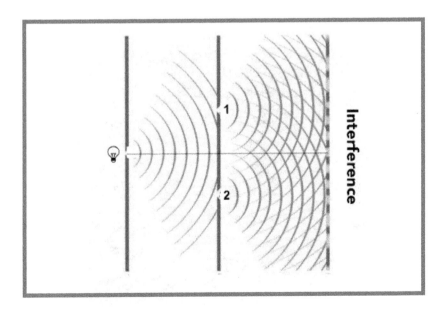

The next image shows it as "cutting," i.e., highlighting the overlap and cancellation of the waves.

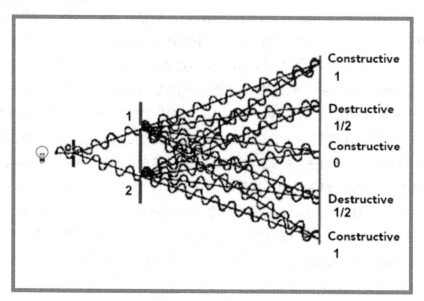

The following image is a photograph that perfectly illustrates the phenomenon in its entirety.

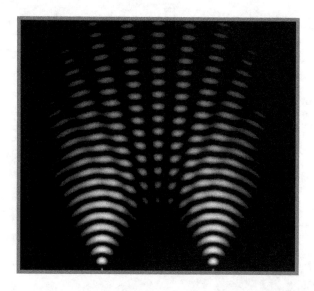

Interference fringes consist of parallel strips of light: light bands are the areas where light waves add up to each other, while dark bands are the areas where they act destructively on each other. Light can be seen by its intensity. The next image shows the interference fringes that originate on the final screen.

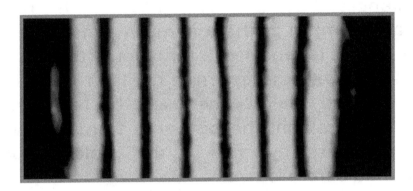

What the fringes undoubtedly demonstrate is that light propagates by waves and not by corpuscles. Interference proves this irrefutably. If only for the simple fact that the light bands are not only found at the holes but both inside and outside them. Only one wave can reach these points, circumventing obstacles, and interfere with another. We will see this essential point again in a moment.

The phenomenon of light interference also makes it possible to easily measure the wavelength, since the distances from the sources (holes) to the point of arrival on the screen are closely related to it. One must consider the distances from the source to the point of arrival to know whether there is constructive or destructive interference.

If the light paths from the holes to the back wall differ by an integer number of wavelengths (0,1,2,3), two amplitude maxima are superimposed, and there is a clear fringe (image below). If, on the other hand, the distances differ by half a wavelength, there is a maximum at the minimum and the total amplitude is canceled. It is not for nothing that the brightest maximum is right on the back wall, exactly halfway between the two holes. In that position, the distances from the two slits are perfectly equal, i.e., the whole number defined before is zero.

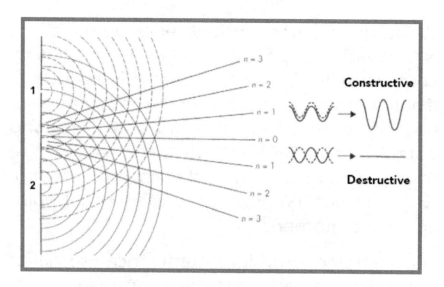

Again, the very fact that the maximum light is not in correspondence of the holes, but in the intermediate point between them, tells us in a very simple way that light is able to circumvent obstacles and, therefore, can only propagate through waves, as in the sea. Just to make sure we don't miss anything, I also show you the next image that summarizes the experiment with a three-dimensional vision.

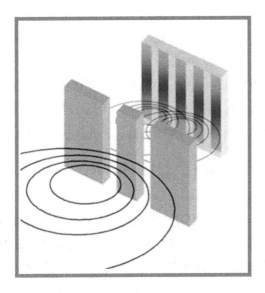

Let's leave the light aside for now, although we'll be back on the interference very soon.

Up to this point, you will have noticed that I have described the phenomena using classical physics. The whole phenomenon not only proves itself but is easily verifiable. Young has made a masterpiece by staying on the tracks of Galilean scientific thought. Precisely for this reason, his conclusion is fantastic, but it is not completely true since it is actually based on the behavior of particles like photons that follow the absurd rules of QM.

I know, you're a little disappointed. A lot of waiting and still not even a trace of QM. Be patient. The important thing is to understand very well the phenomenon of interference in its generality. Only by understanding well what logic shows us, clearly and irrefutably, will we be able to fully reveal the absurdity of certain behaviors that deviate from it. Before facing it, the waves and their characteristics must become an obviousness, a habitual phenomenon like seeing anything that has a minimum weight falling on the ground.

Let's begin, then, the Feynman experiment, one of the greatest masterpieces of the human mind. Pay attention to every single step of the experiment; only then will you understand its wonderful lucidity and essentiality. We will stop at the point where classical physics still manages to provide comprehensive and satisfying explanations, before the final leap towards the absurd. Everything will be easily understandable (Feynman uses simplicity, the best weapon to describe big problems), after having already digested the interference of the waves. However, read very calmly, without taking anything for granted and obvious.

Feynman's basic idea was to compare the behavior followed by single particles and waves under the same conditions as Young's experiment. Isn't that trivial? It's always the same: brilliant ideas always seem obvious... in retrospect. The comparison would have clarified the differences and characteristics of both situations. The normal logic would have accepted with pleasure this result, but it would have been astonished and incredulous in front of the third phase of the experiment, when as single particles very different "objects" would have come into action.

So, let's complicate Young's experiment a bit, leaving the brain free to detach itself, at the right moment, from everyday reality.

CHAPTER - 3

A DRUNKEN SOLDIER

The first part of the experiment uses a soldier, a submachine gun, and his bullets, taking Young's two-slit patterns on an equal footing. The rifle is placed behind a hole made in a wall. The weapon, however, fires randomly in all directions and not only in front of it, along a fixed and determined line. It's probably a rickety old weapon or the soldier's a little drunk. It's better this way, it's just right for us. At a certain distance in front of the first wall, we place another wall, always made of metal and capable of stopping the bullets that fall on you. In this one, we drill the two holes — or better —cracks, 1 and 2. They will only allow the bullets that pass through them to continue. Finally, at an even greater distance, we place another metal wall, covered with some material similar to sand that will trap the bullet that reaches you. This allows you to collect the bullets that hit it and count them after firing a volley.

Obviously, the bullets come out of the rifle one at a time and so they can't bump into each other during the journey, nor can they break or split into several pieces. Consequently, the bullets that arrive on the sand do so one at a time and are perfectly intact. It seems obvious, but keep it in mind. The following image represents the configuration of our experiment. This figure and the following 12 and 14 are rather approximate, though more than enough to define the problem. I use them because they are Feynman's sketches. To the wonder is added a little bit of emotion.

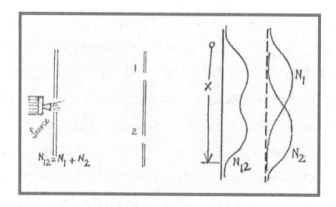

Let the soldier go. Let's imagine that the volley was very long and therefore fired a very high number of bullets (large numbers are fundamental to talk about probability). We count the bullets in the sand according to their position. Let's do something even more mathematically refined. We calculate the percentage of bullets arrived in different positions. In this way, we immediately have the *probability that it has a certain position of the plate to be reached by a bullet.* We could do the same even more precisely. Put a "box" or a "jar" full of sand in different positions of the final wall and shoot 100 or 1000 or 10000 bullets each time. Count how many came into the box and determine the probability for the box in that position. Then we would slide the box to another point (continuously and completely) and off with another 100 or 1000 or 10000 rounds. In the end, we had again the probability of each point on the wall being hit with the same number of shots fired. In short, the box can be considered a bullet detector and, therefore, of probability.

In one way or another, we can easily draw the probability curve as a function of a certain x-coordinate, which indicates the position along with the collection plate. However, bullets or probabilities, the final wall would look like the following image, where the dark stripes are the two most affected areas. At most two and no more than two.

Someone might say to me, "Why not count the bullets without doing too many math steps?" The answer will be obvious soon enough. We need to talk in terms of probabilities since the bullets we will use later have a much stranger behavior than bullets.

The probability curve has a shape similar to the one marked N12 (see image above). Of course, if the soldier aimed accurately, the bullets would be almost exactly behind the two slits. Instead, he's drunk, he shoots randomly and the bullets come in at an angle or touch the walls of the slits and bounce off. In other words, they can fall into the box even when it is displaced from the cracks, albeit in smaller numbers. It follows that the probability line describes a curve that has two peaks at the two slits (more or less) and then fades. This curve depends a lot on the distance between the two slits, as we will see in a moment.

What does the **N12** probability curve represent? Surely it is the sum of the bullets entered through 1 and those entered through 2, expressed as a probability. It's obvious, thinking that each bullet is a single, perfectly defined object. In other words, *the bullet is visible and locatable all along its path*. Keep this in mind, please.

To be sure, however, just do a very simple test

(it is better to be sure of what we do because soon appearances will deceive us!).

Let's close slit 2 and redo the whole experiment. We'll get the **N1** probability curve. Then we close slit 1 and start all over again. We'll get the **N2** curve. If the number of hits is high enough, the statistic assures us that N12 is exactly the sum of N1 plus N2. As already said, each bullet is a single object, and therefore, their total number is obtained by adding up the bullets that arrive on the box, regardless of whether they come from 1 or 2. This fact affects the probability evenly.

Better to be even simpler. If we fire a hundred rounds with only slit 1 open and place the box with the sand in a certain position, we check, for example, that only 10 bullets have hit it. The probability is then given by 10/100. We close 1 and open 2, but keep the sandbox in the same position. This time, only 4 bullets out of every 100 fired will reach the box through 2. That means a probability of 4/100. We can, therefore, say that the probability of having a bullet in the box is 10% through slit 1 and 4% through slit 2. Let's open them both now and fire 100 rounds again with the box (emptied) always in the same position. It is obvious that 10 bullets will arrive on the box through slot 1 and only 4 through slot 2. In total, we'll have 14 bullets, out of 100 fired, in the box. The probability is then

14%, just equal to the sum of the probabilities obtained by opening the two slits individually.

In summary, the probability of bullets coming from both slits is equal to the sum of the probabilities of each slit. It follows, therefore, that the maximum value of the sum of the probabilities of the single slits (i.e., the maximum probability with both slits open) can also occur at the midpoint between them if they are very close together. We see it in the image below, for various positions.

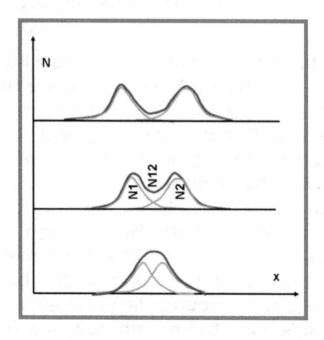

Anyway, this is a decidedly secondary problem that I only touched on so as not to create a completely useless and very explainable confusion: there will be much more important ones. The maximum probabilities can be two or become one, but this depends only on the distance between the cracks and not on the behavior of the bullets. Be careful, therefore, not to confuse this result with the light interference fringe which is always in the middle of the two slits. The really important thing in the case of bullets is that the *sum curve always remains equal to the sum of the individual curves.*

This result seems obvious and even "stupid" to all of us accustomed to living everyday reality. However, it describes a phenomenon that will soon become much more important: there was no interference, i.e., the bullets did not disturb each other (on the other hand they could not do so, according to our logic). In the experiment that has just been done, the bullets come out, in fact, one at a time and cannot bump into each other. Their fate is independent of what the others do. How important is the word "independent" and how connected it is to our logical view of the world.

Let me write a trivial formula, which will be useful to compare it with the (very few) subsequent ones.

$N12 = N1 + N2$ -> no interference

Let's move on to the sea waves. We have already talked about this situation, but it is not bad to repeat it both to not interrupt the logical sequence devised by Feynman and because it never hurts to go back on certain concepts.

CHAPTER - 4

A PERFORATED SWIMMING POOL

The source now is no longer the rifle, but a large mass of water, pressing behind the first wall. In that wall, there is a hole, from which the water manages to pass and flows over the wall. It does this by producing a wave composed of crests and troughs (maximum and minimum). To cause a wave, just shake the water before the wall: the more it is shaken the more the wave increases. To be more exact, a train of waves originates, one behind the other, separated from each other by what is called a wavelength (distance between two crests or between two troughs).

At a certain distance, we place the wall again with the two narrow slits, as in the case of bullets. The wave that has formed after the first opening hits the second wall and manages to cross it only where there are the two slits. What happens at this point? The wave, passing through the slits, splits into two new waves that maintain the same length as the original one. In other words, it is like saying that, if the opening is small enough, it gives rise to a new wave that propagates from it, as if a stone had been thrown at that very point. We have already talked about this behavior concerning light. Actually, instead of water, we could consider light waves (as Young did) and we would get the same result, at least apparently and at first glance.

Let's go back to the two waves that originated in the two cracks. They continue towards the third wall where we had previously put sand for the bullets. Now, we have to change the detector. Let's choose, for example, a cork. When the wave reaches the last wall, it rises and falls following the crests and troughs. The experiment can be seen in the image below. As we did before, with the box full of sand, we move the cork along the whole wall and see how the height it can reach varies. *What interests us is the energy that is released from the wave to the cap or — if you prefer — its intensity.*

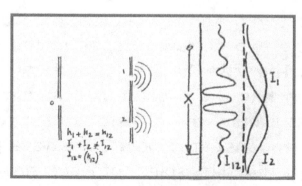

Be careful not to confuse the *amplitude* of a wave with its *intensity*. In the case of the rifle, the energy released and the bullet were practically the same since each bullet released the same energy. For waves, the situation is slightly more complicated. It can be demonstrated (I don't do this because we would have to do a lot of mathematical calculations, but believe me!) *that the intensity is proportional to the square of the height.* On the other hand, we are interested in the intensity or energy released, to compare it, for example, with that of light or bullets.

The final curve, depending on the position of the cap along the wall, is much stranger than the N12 in the previous case. What you see is a figure that also looks like a wave, a continuous up and down, with no relation to the position of the two slits. Precisely, it is dependent on the distance between them, in the sense that the oscillation that can be seen on the wall can tighten or widen. The striped structure, however, does not change. The result (we know it well) is nothing but the phenomenon of wave interference. It is extremely similar to that obtained by Young with light. On the other hand, Young said that light propagated by waves precisely because its behavior was the same as that of sea waves. This second phase of the experiment is also of an almost disconcerting and again obvious banality. But... is Feynman really a genius? Wait, wait, wait...

We still remember that in the case of bullets there were few choices: either it came or it didn't. In the case of the wave, the energy or intensity of the wave can assume all the values between a maximum and zero (flat water).

We call the final curve, which describes the interference pattern, I12. This time, however, to obtain the I12 curve, we cannot add the energies released to the cap with the water passing through a single open slit, as was the case with bullets. To test it, first plug 2 and then 1. We obtain the curves I1 and I2 separately, curves that resemble those of the bullets fired from the rifle through a single opening. This is obvious (remaining in classical physics) since if the wave passes through only one slit it cannot interfere with another and therefore has a behavior apparently similar to that of bullets: a peak that dampens, moving away from the central direction, as the wave loses intensity. The maximum intensity shows it right where it "hits" the wall for the first time. In the case of interference, we have, instead, several maxima and minima (just consider any line parallel to the wall and you can see that the maxima and minima are multiple). The image below illustrates the situation quite well.

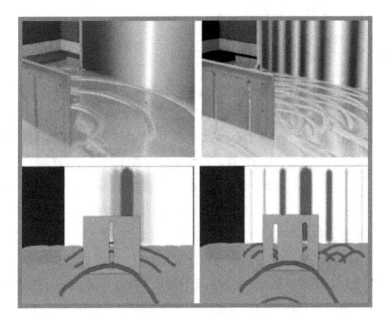

The only substantial difference between projectiles passing through a single slit and wave coming from a single opening is that the projectiles, or -better- the energy released or, if you want, the intensity of the projectiles, is practically their number, in the case of the *wave the intensity is proportional to the square of the height*, that is $I^1 = h_1^2$ **and** $I_2 = h_2^2$. We have to compare *heights* and the *number* of *bullets* since these are the quantities that we detect and that indicate the final probability.

However, we know that the heights of the waves add up and the relationship h1 + h2 = h12 is valid. For the intensity, i.e., for the energy given to the cap, with both slits open, we therefore have (the intensity goes with the square of the height):

$I_{12} = h_{12}{}^2 = (h_1 + h_2)^2$ and, certainly:

$I_{12} \neq I_1 + I_2$ to interference

We need to think carefully about this conclusion which, in practice, provides the best description of the difference between waves and particles. The latter do not show any interference phenomenon *and the probability of finding a particle at a certain point, with both slits open, is exactly the sum of the probabilities of finding a particle at that same point by closing one slit at a time ($N_{12} = N_1 + N_2$).* The waves instead show a different final result. The resulting wave intensity, with open slits, is not the sum of the intensities with alternately closed slits. It is not difficult to consider the intensity as the probability of the cap rising more or less. Think about it for a moment and it will become obvious.

This experiment, which we did with water, is pretty much the same as Young did with light. He had found the interference fringes and could conclude that light could not be transmitted through particles (bullets) since it caused interference. Feynman, for now, only gave a "mathematical" and conceptual definition of the difference between waves and particles hitting a wall. So far, nothing special... it would seem.

We are now entering the fundamental part of

Feynman's experiment and, therefore, I will try to describe it in various aspects to conveniently prepare ourselves for the philosophical speculations that follow. I will be very redundant and repetitive, but each sentence must be digested with calm and extreme attention.

Let the children help you!

In the end, there will still be doubts, however, we will try to fix it.

CHAPTER - 5

A VERY SPECIAL GUN

ere we are finally at the third part of Feynman's experiment, the one that opens the door to QM and shakes (or rather destroys) our ability to understand. To get through this door we have to redo the experiment of the two slits with special bullets, the electrons. We can be sure about them: they are particles that "rotate" around the atomic nucleus, but, above all, they possess a *measurable mass*. Bullets to all intents and purposes, even if they are very small. The source is a filament, the walls are made of tungsten and the detector is an electrical system that can signal every electron that arrives. You could also use photons, but in Feynman's time, it was not so easy to build the right equipment. No, don't be surprised if I mentioned photons after just stating that light propagates through waves. It's part of the game, and you have to accept a first apparent absurdity. Anyway, let's move on to electrons that pose fewer practical problems.

First of all, let us reiterate once again that what we receive in the last wall are indisputably single signals due to single masses. Each signal, a kind of "click," has a certain size (or intensity), always the same. If the source weakens, we only slow down the production of the "clicks," but the single intensity always remains the same, just like the bullets of the rifle. We set things up so that we can be sure that the electrons start, arrive, and are detected one at a time. At this point, we can repeat the experiment done with the bullets and measure the probability of arrival in the various positions of the final wall, as represented in the following image.

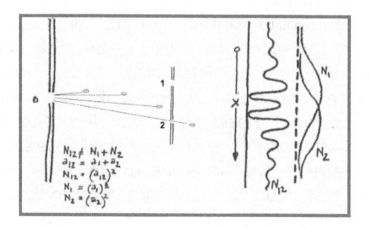

What do you get in the end? Obviously, we would expect the same thing obtained with the machine gun, a curve like the N12 (remember?). But no, absolutely not. What you get is a series of highs and lows quite similar to those due to the interference of the waves produced on the water. The only difference is that the up and down curve does not represent the energy or the intensity of the wave, *but the probability that an electron has of hitting the wall at a certain point of it,* in a perfectly similar way to bullets.

The mathematics describing the curve is very simple and completely different from that of bullets. Instead, it resembles the waves perfectly. It is enough to change the I've written in the case of the waves with N. We know, however, that It is equal to the square of the amplitude h of the interfering waves. Now, what does h represent? We don't know at all and let's indicate it with a lower-case letter for the moment. We can also call it, by analogy, the amplitude of probability, but its true meaning remains a mystery. Just as a real mystery is the entire interference figure obtained by firing "concrete" bullets one at a time.

Did we do real science (in the traditional sense)? Actually no: we just imitated the behavior of the waves since the result is the same. Same aspect and same mathematics. Anyway, to make sure we don't make interpretative errors, let's try to close holes 1 and 2, one at a time, in order to get the real distributions N1 and N2. A perfectly logical attempt, since it is sure that if an electron passes from 1 it cannot pass from 2. One at a time, for goodness sake! Or, at least, let's perform the two experiments, with a single open hole, admitting this obvious deduction (our brain can only operate like this). We close the hole 2 and measure how many electrons send their signal to the final wall. They are only those that have passed through hole 1 by definition and the curve that describes their probability is N1. Then we close hole 1 and draw N2. We could only expect this: the sum of N1 and N2 is not equal to N12. Oh, man! Although the two single curves N1 and N2 are perfectly similar to those obtained with rifle bullets, the total curve is definitely different, identical to the wave curve. We have to be strong and accept the result.

Since **N12 ≠ N1 + N2** there was interference. Yes, but of what? We are more than sure that we used bullets, even if small, and instead, the result seems to be that of the waves. We can only conclude that there is interference and that the probability that an electron arrives at a certain point of the plate is given by **N12 = (a1 + a2)²**, where **a1** and **a2** are two amplitudes. Yes, but amplitudes that make no sense since they are not connected to **N1** and **N2**, unless we impose, again by analogy, that **N1 = a1² and N2 = a2²**.

Let's go over the concept again. The inconceivable thing is that we know very well what electrons passing through hole 1 (they distribute according to N1) and those passing through hole 2 (they distribute according to N2) do. This extremely logical evidence would make us deduce an equally logical conclusion: the N12 distribution, with both holes open, could not be anything other than N1 + N2 as for bullets since there is no alternative: either an electron passes through hole 1 or passes through hole 2. And instead, we have that N12 ≠ N1 + N2, i.e., we have interference.

Let's think about it a little more and repeat the experiment in a simplified way, it may be that we are making a mistake and we don't realize it. Let's also use the image below which is a little less "approximate" than the image above. Let's put our electron detector at a certain point P on the wall. We leave hole 1 open and shoot the electrons. Nothing, the final probability of having any electrons at that point is practically zero. Let's do the same with 2. The result doesn't change. Let's try now, with a little shaking to open both holes. Nothing, still little or nothing. Thank God, it's gone, we made a mistake!

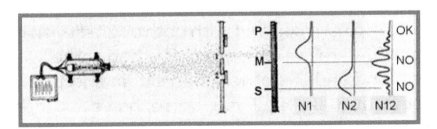

image 15

To be sure, we now place the detector in **M**, right in the middle of the wall halfway between 1 and 2. Still, nothing with 1 open and 2 closed; nothing even with 2 open and one closed. In other words, no electron can reach M, whether it passed from 1 or 2. One long breath and we open both holes. Holy crap! The number of electrons that arrived in M has become enormous. We're right in one of the probability maxima. There's no mistake, it's our brain that's no longer working or it's our senses that are betraying us.

Let's try **S**. We should find a fairly high probability and, instead, the total probability is relatively modest. To get an even clearer idea, let's see the following image.

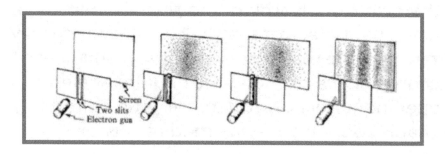

You can clearly see that with both open slits many electrons arrived where very few had come both with one and the other open slit (in short, 2 + 2 does not make 4!). Even worse, however, there were very few electrons where even an open slit had made a lot of them pass through.

It is absurd, isn't it?

Wait a minute, wait a minute. There's some trick down here and probably the electrons went through a hole we hadn't seen or, better yet, they went back and forth. Something similar to the three-card game: it seems impossible, but once the trick is discovered everything becomes trivial.

Wrong! We could turn all the equipment around and we'd see that the experiment wasn't spoiled by any tricks. We just have to think that for some (still unknown) reason the electron broke in two and passed through both holes at the same time. No, nothing to do, this is not the case (at least according to our usual way of thinking).

There is, in fact, no scientifically sound explanation for the final result. We must accept an absurd truth and describe it as such.

Pay attention: Now is the time to NOT use your brain in the usual way. We do not try to explain the phenomenon or find logical errors in the treatment. Let's accept the conclusion without making unnecessary brain effort.

However, we can very well predict the final result, albeit meaningless. This means that, even without understanding it, we could use it for practical purposes. For example, putting something that is "hungry" for electrons in **M** (Image 15) we would be sure to give it the food it wants, even if our logic would refuse to accept it. In reality, this happens now daily with the misunderstood but predictable results of QM.

Before we go any further, let me summarize once again what we have achieved.

The electrons arrive on the final wall one at a time and are forced to pass through either one or the other hole. They act just like the bullets in the rifle. This result is confirmed when the two holes are closed alternately, and the probability of arrival on the wall is calculated. However, the probability that electrons have of arriving at a point on the same wall when the holes are both open, follows the same wave rule. The probability is similar to the intensity of the waves and is determined by squared the amplitude of a "something" that must be the amplitude of a wave, but that we do not know what it is. We have called it the "amplitude of probability."

All we have to do is accept a fact that can only embarrass us and upset our minds: electrons are mysterious and absurd objects, sometimes they behave like particles and sometimes like waves. However, let's be very careful with this sentence, which would seem to partly solve the headache that came while reading. Let's remember that we are sending electrons one at a time and not in pairs, and therefore, we do not have two waves originating in the two holes at the same time as in the case of water. We are forced to keep the headache and use the previous description even if it is certainly artificial and simplistic. I summarize mathematically the whole theory with the following table that will be useful in the future.

Bullets	Waves	Electrones
They come in definite pieces. The body arrives or doesn't arrive	They come in any size	They come in definite pieces. The body arrives or doesn't arrive
The probability of arrival is measured	The wave intensity is measured	The probability of arrival is measured
N12 = N1 + N2	I12 ≠ I1 + I2	N12 ≠ N1 + N2
No interference	Interference	Interference
OK	OK	???

We also have to accept a further absurdity that we had already mentioned at the beginning of the third part of Feynman's experiment: the same thing that can be achieved with photons, precisely those particles that Young seemed to have erased from physics. If the behavior of electrons and photons is so bizarre, what happens to the other elementary particles?

Unfortunately, the same thing. Gosh, we're made of just these particles, but we don't have to follow weirdness like that. When we go through a door, we go through that door and always get to one place by walking or running. We absolutely cannot become waves or remain more or less solid but very concrete particles. Well... that's the apparent luck of complex systems... as we'll see further on.

The experiment would end here and it is summarized in the image below. One could also write many formulas describing the behavior of electrons and the probability of arriving at a certain point under different circumstances. We could also do it and stop thinking about those stupid particle-waves. We know what we get, we can quantify it with "our" mathematics and even apply it to our daily machines and needs. Think that even the Sun does that when it can melt hydrogen and get helium. A man, however, is not (or perhaps he wasn't...) capable of experiencing and accepting what is not logical. And from this nonsense was born the MQ, or, better still, Feynman's experiment contains, for the first time, all the essential points.

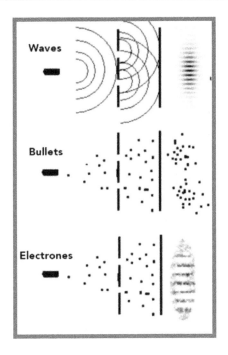

At this point, we would only have to comment on the experiment by making scientific philosophy and also recall some much more recent and sophisticated experiments that have further confirmed the absurdities and their descriptions. Not only that, but they have also brought evidence of something, which we could call "information," that pervades the Universe and that could travel at speeds *well above* the speed of light.

Then we would finally move on to the description of all the laws and rules (if any) of the QM, knowing by now who we are dealing with.

>> CHAPTER - 6 >>

IS IT BETTER OFF NOT KNOWING?

 would like to do something special to make the result even simpler and clearer. Let's be big heads, like those who don't believe in something if they cannot touch it. Time lost, you will see, but it can help us to enter better into the new science. Let's ask ourselves, then, "Are we really sure of everything we have achieved and the very strange phrases we have invented?" Well... it's better to try one more time, trying not to be fooled by appearances. What do you say?

To try to describe all the implications of the Feynman experiment, you have to move step by step without taking anything for granted. We will get to the end and then go back without being afraid of getting bored or feeling like idiots. Each time we'll try to see things from slightly different angles. We have said that our brain has to gradually get used to not rejecting situations outside of any "normal"

logic, but first, it wants to be really sure.

For example, we could say, according to current logic, that the proposition: "An electron either passes through the first hole or passes through the second" is a fact, which cannot be contested. We cannot think that something different will happen since electrons behave exactly like bullets and we have proved this by hearing their "clicks" on the plate. We can call this proposition evidence **A**. Let us then try to discuss it in-depth since it is precisely what seems logical that should raise our doubts. Since the final result is illogical, the leap from common knowledge to absurd reality can be made at any point. Nothing must, therefore, be judged to be safe.

Evidence **A** would suggest that the final sum must be the sum of the electrons passing through the individual holes. Scientific logic dictates this deduction. However, the experiment with the two open holes contradicts the expected result and we must make an effort and admit that proposition **A** is false. It is a very hard choice and against all our deductive capacities.

At this point, all we have to do is admit that it is not true that an electron has to pass through hole 1 or hole 2 by force. Perhaps Alice wouldn't be as surprised as we are. In order not to abandon a logic that is hard to die, all we have to do is to demonstrate, in some way sensible, that electrons pass through both holes at the same time. In short, we are trying to find logic in completely illogical evidence. The brain may be able to do this, but experiments are what they are, and they can't invent things that don't happen. We have to rely on them. All we have to do is look directly at whether an electron passes from one or two or both. In other words, we have to watch them go through the holes. That way, we can be really sure what they're doing.

The truth of the experiment cannot be refuted, whatever the outcome.

In short, we want to check Feynman's experiment once more, going deeper. Hope is the last to die and we would be happy to say that the speeches made earlier are a hoax, a trick of the imagination, or a Nobel Prize trick. After all, who guaranteed that the electrons, once in flight, would actually pass through one hole?

To be able to see something, however, you need to *illuminate* it. We are therefore forced to put a light source behind the holes. If it is sufficiently intense, the light is dispersed by the electron and the part that bounces towards us allows us to locate the microscopic "bullet." In these conditions, we can easily verify what happens before the "click" on the wall. There are three possibilities: either the electron passes from 1, or passes from 2, or it somehow splits and passes both 1 and 2. The light source allows us to verify which event is the real one.

Let's start the experiment.

Although nothing seems to surprise us anymore, we must admit, in the end, that all the electrons that clicked were seen to pass either hole 1 or hole 2. In other words, each electron showed a 100% probability of passing through a certain hole rather than the other. In no case has an electron been seen passing both 1 and 2 at the same time.

Impossible!

We must redo the experiment as a child could do. Perhaps children are closer to QM since they have not yet been affected by the daily phenomenology that is now too deep-rooted in all of us.

Let's get a notebook and make two columns. In the first one, we mark the electrons that have passed from 1, and in the second the electrons that have passed from 2. Let's check, now, their distribution on the wall. Those that passed through hole 1 describe exactly the **N1** curve, while those that passed through hole 2 describe exactly the **N2** curve. Wow, that's a logical result. We have proved that the electrons, with both holes open, always pass through either one or the other hole and that when they reach the wall, they follow exactly the distributions we had obtained by closing one hole at a time.

But, then, both working with one hole at a time and with both the result is the same. But where's that weird interference curve gone? Gone into thin air.

Maybe we drank like the soldier...

Let's even try (even if it would seem a completely useless effort) to draw the distribution of electrons on the wall without making a difference between those coming from 1 and 2. The usual logic would tell us that they must necessarily follow the curve N1 + N2. It's a very critical moment since that's what we did in the previous article with both holes open and that led us to that absurd interference figure. Wonder of wonders! The curve we see is just N1 + N2. Forget interference... We really had too much to drink, we can write that N12 = N1 + N2, just like the bullets in the rifle.

Sorry, let's throw out everything we wrote before: it was just an illusion. Maybe Feynman had been drinking too and with him all the scientists who were influenced by his hangover.

Wait a minute. Before we take action, let's make a quick summary of what we got before and what we got just now.

First, we had closed 2, forced the electrons to pass through hole 1, and found N1. Then we had closed 2, forced the electrons to pass through hole 2 only, and found N2. Then we had opened both holes and found a very different N12 curve from N1 + N2, a typical interference curve. Not happy with this absurd result, we repeated the experiment, leaving both holes open, but checking, directly in "flight," where the single electrons passed from. We saw that all of them, just all of them, passed from 1 or 2. Result confirmed by the N1 and N2 distributions on the wall. Finally, we found, with renewed hope in our normal logic, that N12 = N1 + N2. Two opposite results. The last, however, is logical, the first, absurd. It would be easy to conclude that the first was obtained under unreliable conditions.

We could forget the previous experiment and say that everything is perfect: the electrons, rightly so, behave like rifle bullets. But I thought I was perfectly lucid and not drunk. Feynman loved good wine, but to do what he did in his life, he certainly didn't have to be an alcoholic. How about redoing the experiment just to make sure it couldn't be true? It'll just take a second. Let's open both holes and shoot the electrons without looking at them in "flight" anymore because we have directly verified that they either pass from 1 or pass from 2. With a half-smile of satisfaction, we immediately draw the resulting N12 curve.

No, no! We're really feeling dizzy: here's the well-known interference figure again! We measure for safety, but we already know the result: N12 ≠ N1 + N2. Does the MQ exist or does it not? We have only verified that each time we get different results, sometimes logical and sometimes illogical.

We're about to throw it all away and curse Feynman for making us do this "stupid" experiment. What's the trick?

Calm and cool.

CHAPTER - 7

WE ARE NOT LOOKING

There was something different between the classic Feynman experiment and the one we did a short time ago. In the first case, we didn't look at the electrons immediately after passing through the hole, the second time we did. In the third attempt, we again avoided looking at the electrons and the interference returned to show itself.

The absurdity becomes almost obvious in its absurdity and I am almost ashamed to write the conclusion: "The difference found can only depend on the fact that we turned on the light to look at the electrons while they were in flight. If the light is on N12 = N1 + N2, if the light is off N12 ≠ N1 + N2. In more technical terms we can say that *light affects the characteristics of electrons.*"

Let's use our logical approach once again: light has slightly altered the motion of the electrons and instead of making the electron arrive in the maximum of the probability curve, it has made it arrive in the minimum, destroying the interference figure. We certainly do not solve the absurdity of the interference fringes obtained with the light switched off, but at least we have logically solved the difference between the results obtained with the light switched on and the light switched off.

We try to convince ourselves more and more: electrons are delicate corpuscles and are not as strong as the bullets of a rifle. They are also sensitive to light radiation, while bullets don't even notice it. In this conclusion, which is certainly not the final solution, we already find some truth of fundamental importance: *we cannot use the same laws for macroscopic and microscopic bodies.* Let us keep this in mind.

Let's now suppose we lower and lower the light that illuminates the electrons. Sooner or later, it will be so weak that it will not be able to deflect all the electrons to such an extent that it cancels all the interference fringes. By lowering the light, we should slowly approach the case where N12 becomes different from N1 + N2. Unfortunately, light (we have to accept this because the double-slit experiment can also be done with photons instead of electrons, remember?) does not travel like the waves of the sea, it too moves through particles like electrons.

As I turn the light down, all I do is send fewer and fewer photons to the electrons. This situation is very risky: some electrons could pass undisturbed through the hole without being hit by a photon and we would not see it.

That electron wouldn't be deflected, but our experiment would fail. In other words, less light does not mean less disturbance to the electrons but not seeing them all as they pass through the holes.

We should add a third column in our notebook where we should write: "I didn't see it," every time a "click" on the wall does not correspond to a direct view of the electron through the hole. The weaker the light becomes, the fewer photons are shot and the fewer electrons are hit, and the more the third column becomes full of cases. You can easily imagine the final result: if I add up the electrons of the first and second column they are distributed according to N1 + N2. Those in the third column are distributed according to the strange N12 curve that shows interference.

At this point, it would appear that the matter is settled. Light disturbs the electrons and destroys the interference curve. I don't know how to build it yet, but I know very well how to destroy it. An obvious, yet very interesting conclusion, which it is good to repeat once again: "I cannot directly observe a phenomenon related to microscopic particles without upsetting the same phenomenon." Somehow it seems that we have managed to give a partially logical implication to the whole experiment.

Apart from the fact that this "blurred" conclusion does not explain the figure of interference when the light is off, even when the light is on it gives a distorted view of reality. After Feynman, as he had predicted, much more sophisticated experiments were done to see electrons without disturbing them. However, the result has always been the same. The very fact of knowing where an electron passes disturbs its movement. This is a much more sensible phrase in its nonsense.

In practice, all it takes is the "threat" of a measure to change the state of the system! This "funny" word, "threat," has been proved and confirmed by many subsequent experiments, really amazing, which we will mention later.

Let me, finally, write a few sentences that could describe what is happening and that are now quite acceptable to our brain, which is beginning to adapt to the absurd. We will repeat the concept several times, with small nuances of difference. Never as now has the repetition served to... not understand and, therefore, to enter the world of QM.

Do we decide the reality?

Let's imagine, even if we still don't know how it is possible, that an electron (or a photon) passes through the two holes like a wave and can, therefore, make interference with itself when it hits the wall and be revealed as a particle. However, if we observe the same electron immediately after the holes we can locate it before the final impact and it immediately becomes a particle canceling the interference. This result is not due to the physical disturbance made on the electron, but to the very fact that we have determined its real position in space. Having proved or discovered (or whatever you want) that it passed through hole 1 or 2 we have canceled its possibility of behaving like a wave and it becomes to all effects a particle with the consequences of the case.

In other words, we can "decide" whether to observe the electron as a particle or to allow it to act as an interference wave.

But how can this happen? Have we really entered Wonderland?

Actually, as long as the electron is not revealed on the target, *it is never in a precise point of space* but exists in a probabilistic potential state, described by a wave function, which propagates as a wave and not according to a defined trajectory. *The famous probability wave,* remember? Now you see why we introduced it.

Nothing to say. It's fascinating but increasingly absurd.

Wait a minute, don't hit me... Everything seems finally clear in its incomprehensibility, but I invite you to repeat once again the experiment by changing a little bit the rules of the game in order to better verify this headache hypothesis. Let's remember that our body (and also our brain) is full of electrons (damn!), and yet they are far from being a bit waves and a bit particles.

Twin brothers

We said that we reveal the electron "right after" that passed through the first hole. "Right after" means that very little time has passed since it passed through the hole. But however small the time elapsed, the electron has already passed through the hole; moreover, until now it has remained a wave because we have not yet revealed it. Therefore, in the meantime, the wave has already crossed the other hole and passed through it. Isn't that right? Then how can the electron be revealed "in one piece" near the first hole? What happens to the wave that just went through the second hole? Does it disappear into thin air? It seems so, but how is that possible?

To clarify this point, the physicist Wheeler used

photons and not electrons (we know by now that the result does not change) and proposed to let the photon pass through both holes, like a wave. In what way? Very simple: by inserting a detector not too far from the first hole, but not too close, in short, just enough to make sure that in the meantime the whole wavefront had already passed through the holes. In practice, you wanted to observe the photon as a particle, certainly after it had passed through both holes like a wave. The experiment was called "Delayed Choice" for this reason.

Don't worry, if we try to do it again, the result won't change. The wave disappears into thin air, as the photon is revealed as a "whole" particle, without any doubt. Yet the wave has certainly passed through the second hole too: if you don't insert the detector, the interference figure is formed (which can be formed only if the wave passes through both holes). So how is this possible?! We can only conclude that the information of the photon "caught on the fact" was immediately transmitted to the wave passing through the other hole and it disappeared.

In short, electrons, photons, and all particles exchange information in real-time, more than the "very slow" speed of light.

That's not really the case, but the example is

very good. The reality is that once again we are trying to provide an objective image of what is happening. Unfortunately, we have to convince ourselves that an objective image is not adequate for reality. There is no point in saying that "the wave has already passed," because only when we measure it can we say that something has happened. Before the measurement, the photon remains in an indefinite state of potentiality or non-objectivity (or, better still, unreality).

Back in time

When we then insert the detector, then we can say with certainty that the photon has passed only through the first hole and not through the second hole, and in fact, there is no interference. If instead, we do not insert the detector and we wait to see the photons only when they arrive on the target (with the relative interference figure), then we can say that each photon has made interference as if it were a wave passed through both holes; but we can say this only after the photon is revealed on the target, *even in a point reachable only by a wave and not by a particle*, that is after the measurement.

The thing that seems incredible to us is that what the photon has decided to do (go through one hole or both) depends on a choice after the transit itself! In fact, the detector is inserted after the wavefront has passed through both holes. As Wheeler says, the "choice" of passing the photon through one hole or both is "delayed," i.e., it happens after the photon has passed! This is madness! A particle changes its "state" of now if something happens to it later!

Even the concept of time, of before and after, seems to lose all security.

Let us return to Feynman's experiment and express the conclusion in a more mathematical and less empirical way. We can do this now, also because mathematics is extremely simple. The probability of any event occurring in an ideal experiment (an experiment in which everything is perfectly specified) is always the square of something. We had called this something *amplitude of probability*. When an event can occur in several alternative ways (in the previous case, through one hole or another), we can say that the various amplitudes of each alternative add up. The final probability will then be the square of this sum (interference, **N12 = (a1 + a2)^2 ≠ N1 + N2 = a1^2+ a2^2**). If instead, the experiment is performed in order to determine exactly which alternative has been chosen, then the final probability simply becomes the sum of the probabilities of each alternative (**N12 = N1 + N2 = a1^2 + a2^2**).

This is a general rule that we can extend to all

the phenomena of Nature and it is also the first mathematical formula that we associate with the QM.

The question we are all asking ourselves is, at this point: "Yes, all right, let's accept this conclusion as well. But, what is the "machine" capable of producing this unquestionably illogical reality?" Nobody knows it. We're in good company. We can, like the greatest scientists, only describe the results.

Scientists can give you a very wide and detailed explanation, i.e., they can show you many examples where when the position of electrons is known, the interference figure is immediately destroyed, not only through the experiment of the two holes. The mathematics that describes experiments and conclusions can become increasingly sophisticated to the point of allowing predictions and applications. It can also introduce complex numbers and any other devilry, but the gist of it does not change: the mystery of how (and why) remains a secret even today, a mystery for our brain accustomed to the laws of macroscopic physics.

A completely random kingdom

Let me say something extremely strange, but very enlightening: "Nature itself never knows which way the electron passes. The moment someone or something manages to identify it, the original situation is immediately broken and the interference is canceled. Nature is forced to make a decision.

Let us remember another observation that has been part of our common thinking for a long time: "The same starting conditions must always produce the same final result." Well, it is no longer true in the QM (in part it had already been questioned by chaos).

We could reproduce the same starting conditions and perform all actions in the same way, but we could never know where the electron will pass. Maybe it doesn't know either since it produces two waves of probability that create interference, i.e., he can be in two places at the same time (or rather, wherever it can go).

An electron and a world war

The repercussions of this absurd reality of Nature have practical implications that could be terrible. Let's imagine building a device like the previous one in which if an electron passed through hole 1 it would trigger a contact that would set off an atomic bomb and start World War III. If instead, it passed through hole 2, peace would be maintained (even though it's not exactly peace what we are living today). The future of the world would be tied only and only by the decision of an electron or - better still - by the fact of knowing which way it has passed. We would have a 50% probability of ending Humanity.

What can I tell you?

"Don't look at the electron and let it pass like a wave! It would cause interference on some wall, but it wouldn't blow up a terrifying conflict."

This is an example that reminds us of Schrödinger's cat, which we'll come back to later.

We have really entered into philosophy and it is better to go back and conclude something more precise to describe (only describe, we cannot do more than that) what happens to our electrons.

Electrons are corpuscles when they travel and we can prove it by looking directly at them or revealing them in a non-invasive way. However, they have innate freedom of choice in where they can be at a certain moment and therefore also at the end of their trajectory. It follows that they can be represented very well through the probability that they have in being at a certain point (and, therefore, in passing from one or the other slit).

More correctly we can speak of probability wave associated with them, in the sense that it describes in probabilistic terms which futures are possible for a certain electron.

If we prefer, *the electrons themselves can be considered as a real wave* (in the probabilistic sense of the term, however. Be careful not to think of it as a wave formed by something; otherwise, we would fall into classical physics and goodbye to QM).

However, we know well the characteristics of waves (whether they are probability waves or water waves): they give rise to interference and the fringes seen on the screen hit by the electrons show the areas where the probability waves add up and subtract. The probability associated with an electron has, in fact, always 50% to pass from one slit or the other. These probability waves are the ones that cause interference. However, we must leave the probability associated with an electron free. If we manage to locate it, its probability wave is immediately destroyed (we know exactly where it is), the electron is immediately transformed into a particle and goodbye to interference.

If, on the other hand, we have only one hole, we do not allow the passage from a slit and the probability wave, which continues to exist, cannot give rise to the fringes since no sibling interferes with it. Whether we look at it and transform it into a particle, or let it travel as a wave to the final wall, interference cannot be created. On the final wall, in any case, the wave becomes a particle and it "clicks." After all, it's obvious: the wall was used to detect it. The final probability of hitting a certain point is identical for both particles and waves, as we now know very well.

In even simpler and more general words: at the moment of observation, a particle that has inherent in itself the probability of being in many places (such as 1 or 2), comes to occupy only one of its possibilities and begins to live a concrete existence. Its "state" has completely changed.

His freedom has been destroyed, as it has been discovered where it is, as when you discover a murderer who could be hiding in a thousand places in the city. When, finally, you succeed in capturing him, all the other possible shelters lose their importance. More technically, the probability of being in the revealed position is 100%.

It's impossible for interference to form with another probability wave. The other wave would have a zero probability. The wave turned into a well-defined bullet because of "our fault" (or something or someone who revealed its secret...). The shocking but also enlightening aspect of these discoveries is that the whole Universe —and ourselves— are formed by particles, the same particles that exist as matter when we observe them and exist as waves of possibilities when we do not observe them.

Let me give you another example related to everyday life. Be careful though, it is not a concrete experience, but a purely mental experience. Let's suppose that a person's mood is bad (C) or good (B) and that the probability of finding him or her in one of these two states is 50 %. The QM we have just learned to know then tells us that the mood of the person in question, at any time of the day, is represented by the overlapping of states C and B, but that the probability P of finding her/him in a bad or good mood is in the ratio 0.5 to 0.5. We indicate with P(C) and P(B) the probabilities of being in the two cases C and B.

This condition can be written symbolically with the trivial formula: $P = 0.5\ P(C) + 0.5\ P(B)$. We can, therefore, also say that the person's mood oscillates (just like a wave) from state C to state B. To find out more, we must meet him/her ("observe him/her") and verify what his/her mood is. We can, for example, find him/her in state B, which automatically allows us to say that at the moment of the encounter we have reduced his/her wave function (probability wave) to only state $P = P(B)$. Very simple, isn't it? This discourse seems more than logical when we talk about states of mind, that is, abstract thoughts. We accept it much less when we refer to solid and concrete particles.

The fundamental difference, concerning the analysis of phenomena, between classical and quantum physics, is that in the former the interaction between objects and measurement devices can be neglected or eliminated, while in the latter this interaction is an integral part of the phenomenon. It is not the interaction with the particle (classical physics), but it is the determination of the exact position that destroys the probability wave that possesses 50% to pass from 1 and 50% to pass from 2. As soon as we know where it is, that position possesses 100% to be true and in all the rest of the space the probability becomes exactly 0%, that is nothing.

Any "microsystem" (photon, electron, etc.) is not obliged by deterministic laws to follow precise trajectories. "Probabilism" explicitly forbids any elementary particle to have a defined trajectory. In the case of the two-slit experiment, even a single electron, or photon, travels all the possible trajectories between the source and the target wall. Going through all the possible trajectories, the particle inevitably meets also the two slits, succeeding, after having crossed them (and having, therefore, interfered with itself), to produce on the screen the light to dark bands typical of constructive and destructive interference.

When the particle passes through slit 1 this

determines a certain possible world (which we will call world 1); when it passes through slit 2, we have world 2 instead. In our case, it means that both these worlds coexist in some way, one superimposed on the other.

Well, I have pronounced several sentences all very similar, but which could better refine the thought and the relative concept (abstract by definition).

That's enough for now and let's continue with something more technological, but really amazing. We should already expect it, but the wonder of the experiment that follows is always indescribable. One of my absolute favorite ones to accompany me to the madness of Alice's wonderland.

CHAPTER - 8

ASPECT'S EXPERIMENT

'll introduce you to two of them, somehow similar. You choose which one you like.

Let's consider, first, the *Aspect experiment* of 1982. It is based on two correlated photons (i.e., with symmetrical characteristics, since they too have a "character," as we will see later) fired simultaneously along with different directions. They too, in the end, arrive on a detector able to count them. Along the path of one of the two, however, can be inserted into a filter that can change the direction.

The fantastic result is that when it deflects the first electron, it immediately deflects the other one too, even though it is more than thirteen meters away. The extraordinary fact, already foreseen and feared by Einstein since 1935, gives clear evidence of a reaction, following a certain action, which takes place practically "in real-time," almost as if there were among the related particles an instantaneous transmission of information that doesn't care about the insuperability of the speed of light (imagine the effects on the principle of simultaneity). Twin photons "feel" the same problem at the same time. Is it likely that monozygotic twins have precisely this type of related particles in their neurons?

For those who would like to know a little more, the image below shows an outline of the equipment used by Aspect. In the center, we have an excited calcium atom, which produces a pair of related photons moving along opposite paths. Along with one of these paths, from time to time and in a completely random way, a "filter" (a birefringent CB crystal) is inserted. Once a photon interacts with it, it can, with a probability of 50%, divert it or let it go on its way undisturbed. A photon detector is placed at the extremes of each expected path.

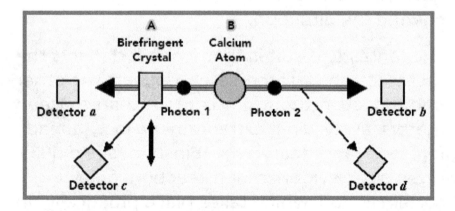

Now, the extraordinary thing verified by Aspect

is that when the birefringent crystal CB is inserted and a deviation of photon 1 towards detector c is produced, even photon 2 (without "obstacles" in front), "spontaneously" and instantaneously, deviates towards detector d. Practically, *the act of inserting CB,* with the consequent deviation of photon 1, instantly and remotely diverts photon 2. I told you, they are like Siamese twins...

What does this inexplicable and astounding experiment tell us? That we must abandon the idea that related particles, located in distant places, represent *distinct entities.* At the same time, a good part of the conceptual obstacles that prevent non-local communication or action also disappear. If we act on something, we get something else even if their interconnection is part of a describable but illogical situation. We can make something happen in place A, acting on place B, without there being any concrete relationship between the two places and between the phenomena that happen there. Ubiquity? Monozygotic twins? Infinite speed? No, just MQ! As the Nobel Prize in Physics Brian Josephson said: *"The universe is not a collection of objects, but an inseparable network of vibrating energy models in which no component has a reality independent of the whole: including the observer in the whole."*

We made our acquaintance with "entangled"

particles in an embrace that can never be forgotten.

Those who would like to ponder a little would understand that this experiment has already solved one of the problems linked to inflation. We have particles (and therefore also macroscopic objects that are composed of them) too far apart to have exchanged information, and yet they behave in the same way. The solution is obvious... don't you think? Don't worry, we'll get back to it.

Do particles read our minds?

The experiment, carried out at the beginning of the 90s by Mandel, is still one step ahead of Aspect's, even if it somehow resembles it. It undoubtedly represents a fantastic, but perfectly reproducible sleight of hand. While we don't know what's actually happening physically, we do know how to use it for technological purposes.

First of all, we recreate a situation similar to that of the photon passing through the two slits, but through a different device, i.e., a semi-reflecting mirror (also called divider): it transmits light at 50%, i.e., only half of the light intensity is able to pass through the mirror, while the other half is reflected. In short, something very similar to the CB of before.

Analyzing the single photons, in a traditional

description, we would say that the probability that a photon crosses the mirror (instead of being reflected) is 50%. If we consider 100 photons, according to the conventional logic, we statistically expect 50 photons to pass through the mirror, while the other 50 are reflected: the initial beam of 100 photons will be, therefore, divided into two different beams running through different paths. This, however, is true only if we can reveal the single photons, otherwise, we have to admit that each photon is in a strange "state of superposition," that is 50% crosses the mirror and 50% is reflected.

In other words, the path of each photon will be indefinite, since "half" passes through the mirror and "the other half" is reflected, although it is indivisible. But we already know these things well enough.

If we do not explicitly measure the path followed by the photon and have the two potential paths engraved on a screen, we get the usual interference figure, i.e., the photon (while remaining a single particle) has passed through both paths and eventually produces interference with itself. So far, it happens what we have already described too many times. The fundamental difference is that this time the mysterious splitting of the single-photon is not caused by the two slits but by the semi-reflective mirror.

As you can see in the figure below, laser 1 emits a photon, the semi-reflective mirror 2 "divides" the photon into two "ghost" parts, each of which travels a different path, 3 and 4. The mirrors in points 3 and 4 are "normal" and serve only to direct the two paths appropriately: the waves remain waves.

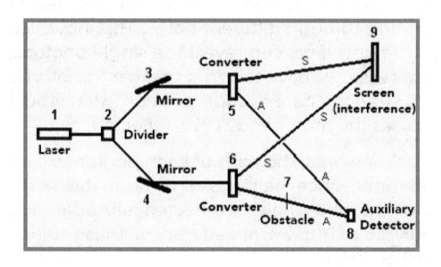

On each path, there is a "converter." Let's not go into details and trust the perfection of the experiment: I guarantee 100%. Each converter, 5 and 6, divides its ghost photon into two "twin" photons of half energy. One is called "signal photon" and is indicated with S, while the other is called "auxiliary photon" and is indicated with A. Finally, the two S paths are revealed on screen 9, while the two A paths are addressed on the auxiliary detector 8.

Let's see how the whole system works: laser

1 fires a single photon at a time that affects the semi-reflective mirror 2. Since we don't measure which path the photon takes, it passes like a wave from both paths 3 and 4, and in converters 5 and 6, it is divided into twin waves of half energy. In the end, the two "signal" paths (indicated by S) affect screen 9 where the photon S will interfere with itself (i.e., with the other part of itself passed by the other path) and finally become a particle (it has been detected).

Then we will fire more photons from the laser, one at a time, and at the end, as a result, we will see a clear interference figure on screen 9. So far, so good as always. We're getting the gist of the experiment. However, let's not forget auxiliary photons A. They too meet in 8, with the obvious consequences. Even there we hear a well-known "click," like the one that occurs in 9.

We were very good, but we did nothing special. We managed to obtain the usual experiment multiplied by two, through the use of mirrors that do not alter the probabilistic wave state by definition. All right, here's the best part. Let's see what happens if you put an obstacle in point 7.

The paths of S and A are now divided and we expect them to be independent: the obstacle in point 7 should not alter the interference figure in screen 9, since point 7 is on another path, even very far away, leading to the auxiliary detector 8 and not to screen 9. In other words, only the interference figure of 8 should be destroyed.

That's not true! If we insert the obstacle in point 7, thus interrupting the path of an auxiliary beam, the interference figure of the "signal" beams S in screen 9 also disappears! Yet we did not measure the "signal" beams (which end up on screen 9), but only an auxiliary beam (which ends up in detector 8). Even if we move the two beams, A and S, very far apart from each other, when we operate on beams A we incredibly affect beams S as well.

How is that possible? What has changed since the previous case when there was no obstacle in point 7? Please be careful what I'm about to say! *The "potential knowledge" we have on the "signal" beams has changed.* Since the path passing through obstacle 7 is interrupted, when we reveal a photon on the auxiliary detector 8, it must necessarily come from the path passing through mirror 3 (it cannot come from the path of mirror 4, since it is interrupted in point 7). Therefore, by measuring its interaction with the signal photon on screen 9 we would be able to say with certainty that that "signal" photon must come from the path of mirror 3, i.e., we would know or — better still — we would deduce that the photon has passed "entirely" from this path and consequently cannot be passed from the path of mirror 4 (by now it has become a particle since we have discovered its trajectory through simple deduction).

It is obvious, therefore, that it cannot interfere (as in the case of the two slits). This explains why the interference figure in screen 9 is destroyed if we put an obstacle 7 on the auxiliary beam.

Wait, do you want a concrete example? You don't need to catch a thief red-handed to arrest him. It's enough that the investigation made in a closed room of the police proved his guilt, through irrefutable evidence. Maybe the thief has run away abroad by now, but his fate is sealed: he is now a thief for all intents and purposes.

The remarkable fact is that it is a disconcerting "action at a distance": by acting on point 7 we alter the physical state in a different place, i.e., on screen 9, where the interference figure is destroyed, and this is only due to the fact that we can deduce which path the photon has followed on screen 9. Direct knowledge is not necessary, but a piece of information or, if you prefer, an act of awareness is sufficient.

This potential knowledge is sufficient to alter the physical state on the signal detector, destroying the interference figure. Someone might say, "Yes, but the scientist or police inspector may not be smart enough to make the deductive reasoning that nails the thief. No, Nature considers us all equally "intelligent," precisely because we are made from the same particles. For the laws of the Universe, there are no ignorant or luminaries, there are only creatures, the creatures it built in the first moments of life. We hope not to make you understand that unfortunately, it is not always true... maybe it would destroy everything you have created so far.

I'm kidding, of course, but I'm not completely sure. After all, even the inability to understand can be considered interference, an anomalous defect and we still do not know how the particle-waves behave in these cases. For those who want to think, it is not difficult to imagine, then, parallel universes or random choices or whatever you prefer. God may be playing dice. Even Einstein can make mistakes...

It seems that quantum mechanics gives a new message about the structure of reality, and sanctions the end of materialistic objectivity, in favor of an "idealistic" conception, in which objects exist in an "abstract" and "ideal" state that remains theoretical until the perception of a conscious subject makes it real. Or, without making it so big, we can limit ourselves to note that the classic materialistic model is inadequate to describe quantum reality and we must turn to models that conceive the Universe in terms of "information" rather than "matter." Ask yourself: "Does the Moon exist even when we do not look at it?." It seems and it is a silly question, but for the QM it's not really like that...

QED

I'm jumping ahead a little bit just to build some kind of bridge. The exploration of the subatomic world has revealed the intrinsically dynamic nature of matter; it has shown that the constituents of the atom, the subatomic particles, are dynamic configurations that do not exist as isolated entities, but as integral parts of an inextricable network of interactions. These interactions involve an incessant flow of energy that manifests itself as an exchange of particles; a dynamic reciprocal action in which particles are created and destroyed in an endless process, in a continuous variation of energy configurations. The interactions between particles (awareness or certainty of an event or whatever you want) give rise to *the stable structures that form the material world*, which in turn does not remain static, but oscillates in rhythmic movements. The entire Universe is therefore engaged in an endless movement and activity, in an incessant cosmic dance of energy.

This dance involves an enormous variety of

configurations but, surprisingly, these fall into a few distinct categories. The study of subatomic particles and their interactions, therefore, reveals the existence of perfect order. Alice's wonderland isn't as crazy as it seems after all. It's just us who can't understand its precise and even simple rules.

We will see that everything about QM has already completely "blossomed" through Feynman's experiment. His experiment, and its natural implications, remain one of the highest points in all of science, not only in the last two centuries but in all of human history. A simple and wonderful summary of a world that no one, he first, has yet been able to understand. Perhaps it will take a little girl, perhaps named Alice, to explain (and not only to describe) the new logic.

Personal thoughts

Allow me a very personal reflection: does objectivity still exist in the QM wonderland?

The macroscopic world around us seems to follow the fundamental rule of cause and effect. In other words, everything that happens is linked to a more or less macroscopic cause. Even chaos is somehow linked to an infinitesimal variation of the initial conditions. In a truly "crude" way, nothing in the world seems to be able to choose its future. Also the animals, after all, through instinct, follow similar rules.

What does the human brain do? It too theoretically has infinite (or almost) choices ahead of it, and only the interaction with something that happens forces it to make a choice. There is a real difference between the possibilities of the mind and those of elementary particles. I would say not. The entity "mind" is an elementary particle, just the one we associate with the diversity between the sentient human being and everything else. What could I conclude then? That the aliens, that is, the characters endowed with a reason and a capacity for thought, are already among us, indeed we are made of them.

Think of what we call a person's character. What does that mean? In simple terms, the difference in action and/or reaction, with equal freedom or possibility, between one brain and another. A combination such that they react differently to the same urges. A wave and nothing else. Well, it's the same for elementary particles. Do they also have their own "character," a certain "spin," a certain "scent," etc.. Brains in freedom, waves of probability, or possibility? Certainly, but linked together by a bond still denied to us (very crude) thinking beings: the ability to inform ourselves in real-time, to act at a distance, to adapt to what our brothers and sisters have "decided" to do. Let's not delude ourselves too much to think that it is us with our experiments and our actions that modify the behavior of elementary particles. Think about it: they command us. Do we really need our consciousness to change the state of a particle?

Well, I'm sorry. I've been doing a little philosophy that may or may not interest the description of the QM. I, in my small way, consider it extremely useful to be able to say — following Feynman's words — "I realized I didn't understand." No fear, though, the elementary particles will teach us... sooner or later!

We will return to these concepts in a little while anyway.

Anyway, the QM we have just begun to know already solves the "historical" problem about the nature of light. When it is not revealed, it behaves like a wave (the interference proves it irrefutably). When it is observed, or even just deduced, it immediately becomes a particle, the photon. And no wonder, since its possibility of being everywhere is canceled. This double "nature," carefully hidden by the microcosm, has been demonstrated wonderfully by Feynman's experiment. At this point, we are truly able to appreciate its genius.

We will see that Planck's constant, the uncertainty principle, the atomic orbitals, the "character" of the particles, the symmetry, and its rupture, the formation of matter, the phase transitions, the quantum vacuum, etc. "hover" more or less hidden between the two fantastic slits.

>> CHAPTER - 9 >>

SCHRÖDINGER'S CAT

When talking about entangled particles and superimposed states of the same particle, reference is often and willingly made to Schrödinger's cat paradox experiment. The great physicist was able to describe a mental experiment (don't worry for the kitten, he never really performed it!) that transferred the superimposition of two states of a particle (in this case an atom) to a macroscopic object, that is a cat. In this way, he confronted his colleagues with a verifiable absurdity in everyday reality.

In reality, Schrödinger had proposed it only to provoke and demonstrate how extending the concept of "overlapping states" required a lot of attention and led to unacceptable solutions. In other words, he who had derived the wave function, the basis of the whole QM, had many doubts about a completely probabilistic and not deterministic view of the world (he was, in short, on Einstein's side).

Often, on the other hand, his "cat" is presented as an experiment that proves just the opposite of what the author himself thought. Common problems in the divulgation of what is not well understood by those who divulge.

It is worth, however, to describe it and use it to summarize many ideas and try to understand how it can be "disassembled" in various ways, without violating the basic rules of QM.

Let's take the exact words of the Austrian physicist, which make his motivations perfectly clear to us:

"One can even set up quite ridiculous cases. A cat is penned up in a steel chamber, along with the following device (which must be secured against direct interference by the cat): in a Geiger counter, there is a tiny bit of radioactive substance, so small, that perhaps in the course of the hour one of the atoms decays, but also, with equal probability, perhaps none; if it happens, the counter tube discharges and through a relay releases a hammer that shatters a small flask of hydrocyanic acid. If one has left this entire system to itself for an hour, one would say that the cat still lives if meanwhile, no atom has decayed. The psi-function of the entire system would express this by having in it the living and dead cat (pardon the expression) mixed or smeared out in equal parts.

It is typical of these cases that an indeterminacy originally restricted to the atomic domain becomes transformed into macroscopic indeterminacy, which can then be resolved by direct observation. That prevents us from so naively accepting as valid a "blurred model" for representing reality. In itself, it would not embody anything unclear or contradictory. There is a difference between a shaky or out-of-focus photograph and a snapshot of clouds and fog banks."

Before I continue, let me make a small and insignificant modification since I am a gutted lover of cats and animals in general. I would prefer to replace the flask of hydrocyanic acid with a nice anesthetic (rightly dosed) that makes the cat safely sleep deeply for at least an hour. Agreed? Ok, let's continue.

The case of the cat is easily attackable and rebuttable. Various ways have been used. For example, we can say that the superimposition of state can happen only in a closed system (somehow as if the chamber represented a Universe separated from ours), while instead, this is not true and through any kind of radiation or very thin interrelations between chamber and external world this does not happen and therefore the wave function collapses immediately and the cat is really either alive or dead even if we cannot "apparently" ascertain it.

Or, we can say that the cat is never in the superimposition of states because it is already broken by the Geiger counter, placed between the radioactive atom and the poison, which serves precisely to release the poison: this is to all intents and purposes an observation.

Let's not forget, however, that still today we think of a much more fascinating solution: the cat is really in two superimposed layers that both exist in two separate universes. In one the cat is alive and in one the cat is dead.

In short, the implications and discussions, not wanted by Schrödinger himself, continue today and lead to truly extraordinary visions and theories even if purely theoretical and not verifiable with "ad hoc" experiments. We remember that the oddities of the microcosm are now perfectly verifiable (as we have already seen partially) and used in the most advanced technology in all fields of science. But not only that. Phenomena like the nuclear fusion inside the sun star (and its numerous sisters) can happen only and only if certain principles of the QM occur. In particular, the tunnel effect, which is linked to *Heisenberg's uncertainty principle*, which really commands all "microscopic" dances.

There remains, however, a profound doubt still unresolved. What is it, and where and how is the boundary between the microscopic and macroscopic world established? The application of the uncertainty principle succeeds in resolving much of this problem. Even stronger is the doubt about the actual need to have a conscious verification of the position of the particle. Is the Geiger counter (devoid of intelligence) enough to transform the wave into a particle and/or a device that does not warn us in any way from which slit the photon has passed or is it really necessary an effective awareness of the phenomenon? On this still open question, we will return shortly in the conclusions.

Let us not forget, finally, that the non-deterministic view of reality is far from resolved. The problem of missing variables raised by Einstein and colleagues seems to regain vigor and a recent macroscopic experiment seems to open new visions of reality, bringing us back to de Broglie's famous pilot wave. A kind of deterministic compromise in which all the description of the QM would remain unchanged, but would be controlled by a kind of material wave that guides the particles, which would never actually acquire the "state" of waves but would adapt to the pilot wave. A very preliminary example? Think you have a liquid that has no evident vibrations, but that is traveled by waves below the detection threshold. Drop a drop of the same liquid. You'll see it disintegrate and move into the liquid pushed only by the wave it caused. The drop remains a drop but takes on the characteristics of a wave.

Before moving on to the preliminary conclusions (remember that this is only a guide for beginners), which summarize and generalize what we have seen so far, let me recommend to the most willing a mind game that requires a great logical and interpretative effort. Remember the famous board games in which situations such as: "One door leads to Hell and one to Paradise. In front of each door, there is a guardian: one always tells the truth, and the other, a lie. We do not know, of course, which of the two is the liar and which is the honest one. What question do we have to ask either of the two guardians to make sure we go to heaven?" Well, raise the mental game to a slightly higher level and you can read this article that shows that to make the cat's wave collapse there is no need to have a direct awareness of the state. It sounds like a pure game, but its conclusion is of great interest.

CONCLUSION

Maybe, indeed certainly, I will repeat myself a little bit, but, as you know by now, this is one of my "strategies" of divulgation, which many people appreciate and, therefore, I keep doing it.

First of all, let me make one observation that is certainly not new to you. The QM was born in the same years as the theory of relativity and was, in a similar way, a reference theory for the whole 20th century. However, it has never really been able to get out of the narrow circle of insiders. One might think that this is due to the mathematical difficulties of the expressions that govern the wave function and not only complex plane and similar things. No, it is not enough to explain its "ghettoization."

There must be something else that seems to preclude its disclosure. Relativity is no less than that, but it has entered overwhelmingly into common language. Moreover, QM is at the basis of all the technological innovations of today, from atomic energy to computer microelectronics, from digital clocks to lasers, semiconductor systems, photoelectric cells, diagnostic and treatment equipment for many diseases. In short, today we can "live" in a "modern" way thanks to MQ and its applications.

Our mind, as I mentioned earlier, seems to be based on quantum processes, including state overlaps, wave collapses, and entanglement situations. The real difficulty lies in its "counterintuitive" postulates about the reality of Nature. A real discomfort in entering an unknown and absurd world like Alice's one. Let's not feel too inferior, though... The founding fathers themselves lived this situation to the limit of the absurd. Could one really believe that Nature followed completely random rules or, instead, was it all an appearance due to the lack of information, of a deterministic kind, still missing?

The very creator of the very general and ultra-confirmed principle of indeterminacy (Heisenberg), said: *"I remember the long discussions with Bohr, which made us stay up late at night and left us in a state of deep depression, not to say real despair. I kept walking alone in the park and I kept thinking that it was impossible that Nature was as absurd as it appeared to us from the experiments."* In a nutshell, there is no defined and describable reality, but an objectively indistinct reality, composed of superimposed states.

Let's pick up on two essential points that we have learned to know, but certainly not to understand:

1. Every action of the finer structure of matter is characterized only and only by its probability of happening. Phenomena completely acausal, not deterministic. But, above all, by the indistinct separation between the observed object, the measuring instrument, and the observer.

2. It is possible that, under certain conditions, what happens in a certain place can drastically influence what happens in a completely different place, instantly. This leads to the phenomenon of entanglement, the twisting of particles that have had an interaction in their past (but, recent research also seems to admit "contacts" in the future) or that were born "together." Although completely separate, they always represent the *same entity*. An action performed on one has an instantaneous effect on the other.

Perhaps you've already noticed the real problem with the QM. On the one hand, the difficulty of dealing with concepts that are too far removed from everyday reality, and on the other hand, the difficulty of using proper language to explain this absurd world. Math can also describe it, but the letters and words of this strange alphabet are missing. Exceptional, in this respect, was Feynman's work with his diagrams applied to QED (which we now know quite well).

It is interesting to quote a sentence by Max Born about this: *"The ultimate origin of the difficulty lies in the fact (or philosophical principle) that we are compelled to use the words of common language when we wish to describe a phenomenon, not by logical or mathematical analysis, but by a picture appealing to the imagination. Common language has grown by everyday experience and can never surpass these limits. Classical physics has restricted itself to the use of concepts of this kind; by analyzing visible motions it has developed two ways of representing them by elementary processes: moving particles and waves. There is no other way of giving a pictorial description of motions – we have to apply it even in the region of atomic processes, where classical physics breaks down."* In a nutshell, the description of the QM itself could be heavily influenced by our "classic" descriptive limits.

Often, therefore, the founding fathers themselves have used analogies and similarities to express purely mathematical concepts. They, however, must be considered for what they are and should not be given any real and concrete validity. This is a really huge problem for our brain (especially today) even if — perhaps — it would have all the basics to use an adequate language, but still too indistinct to be formulated correctly: Feynman's diagrams, I repeat, are a wonderful attempt in that direction.

Niels Bohr himself used graphic analogies to try to support such absurd theories for our classical language. Famous is the white vase representing, at the same time, two black human profiles.

A state of superimposition between two realities existing instantly (two states or — maybe two Universes?). This type of analogy has influenced many optical illusion games and even artistic currents (think of Picasso).

It is a pity that these interpretative efforts,

combined with Feynman's more complete and refined efforts, do not find their way into schools in order to adequately prepare young people to "stammer" their first quantum words and to begin a primitive language that would allow them, today, to understand, at least partially, the reality of Alice. And not only passively undergo the most wonderful technological applications that are now an integral part of their physical body. Real "appendages" which, however, act unconsciously, independently of any mental command. Unconditional reflexes and nothing more.

After all, de Broglie advanced his daring hypothesis precisely by following the symmetries of visible Nature. He only associated with the matter in general, what happened to the light. In short: if light manifests itself under a double aspect, undulating and corpuscular, why not think that also matter follows the same rule? It is enough to associate to each corpuscle of matter a wave of a certain length, that is, a phenomenon extended to the space surrounding the particle. The dualistic nature (particle-waves) applies to all particles, such as electrons, atoms, and other moving entities.

However, the basic problem remains open (still today the subject of discussion and interpretation) which we have mentioned. The wave of matter that commands the particle can be deterministic, and therefore still unknown in its real structure (in line with Einstein's idea) or, instead, a different representation of the same particle and therefore follow the rules of complete acausality (Copenhagen school).

In one way or another, however, it must be concluded that the light or a beam of electrons is nothing but a "train" of electromagnetic waves, but also a jet of "bullets" as in the double-slit experiment. While remaining in this basic ambiguity, Schrödinger formulated the equation that perfectly describes every undulatory property of matter through its wave function. It allows us to describe every single behavior and, above all, to calculate the probability distribution to find a particle inside the associated wave. Overwhelming mathematics that, however, does not annul the fact that Schrödinger himself did not believe in the actual concreteness of this representation. Everything and the opposite of everything (conceptually), but described in the same way.

However, its equation clearly confirms what Feynman's experiment illustrated above: a particle can occupy ALL possible positions within the associated wave. By occupying all possible positions, it no longer has an actual place of existence or direction. It automatically cancels any possible prediction of its future except in purely probabilistic terms (the QED is increasingly understandable... don't you think?). The pilot wave or a hidden variable does not change the action of Nature and its probabilistic description.

Once again we fall back on Heisenberg's principle... In classical mechanics, deterministic essence automatically allows you to predict the future if you have exact information about the position and speed of a particle. Let us remember, in this regard, that the first mathematical methods that allowed the calculation of an orbit of a "planetary" particle were based (and still are based) on the knowledge of at least three positions and three velocities, such as to allow the solution of an orbit characterized by six unknowns. Too easy for microscopic particles.

The probabilistic conception leads inexorably to the principle of indeterminacy, inherent in the whole microcosm: either one knows the position or one knows the speed. To know both with accuracy is impossible, otherwise, the particle would be located and the wave would collapse. And we go back to the starting point again. Whether there is an initial causality (completely unknown) or not at all. In a nutshell, the double-slit experiment perfectly illustrates all the problems of QM.

It's worth reflecting on Einstein's dramatic emotional situation. While he was giving physical reality a perfectly deterministic representation, he found himself involved in a representation that led to the complete acausality of Nature. He said: *"Quantum radiation theories interest me very much, but I wouldn't want to be forced to abandon narrow causality without trying to defend it to the limit. I find the idea quite intolerable that an electron exposed to radiation should choose of its own free will not only its moment to jump off but its direction. In that case, I would rather be a cobbler, or even an employee in a gaming house, than a physicist."*

Yet, no physicist has contributed as much as Einstein to the creation of Quantum Physics... What he demonstrated about it (and for it) was enough and advanced for a scientific career of the highest level (not for nothing earned him the Nobel Prize). It is, therefore, easy to understand his existential drama, which never left him until his death. A mixture of anger, wounded pride, unshakable confidence, and despair at not being able to demonstrate his certainties.

This mix of frustration, exaltation, hope, disappointment, innovation, and conservatism has permeated all the great minds that gave birth to QM. A very choral work and certainly not a puzzle of individual ideas. Each one, almost unwillingly (sometimes even against their purposes), did nothing but put an extra brick to a building that was becoming an incredible skyscraper with an increasingly solid, unassailable foundation.

Perhaps this very unique way in the history of science to formulate a more and more complete and refined theory, by many higher minds, could make people understand that QM is something really inherent in the human mind, but that it has extreme difficulty in coming out of the closet. Surely, the knowledge of the language of Classical Physics has made enormous steps forward, but not so far from the almost unconscious insights of Democritus and Epicurus. In a nutshell, the mind must be trained to indulge in a reality that is only historically and culturally absurd.

The more we get into the very essence of QM and its principles, the more fundamental and complete the double-slit experiment becomes. A true scientific masterpiece, a manifesto itself of the future of the human intellect.

Let us recall the basic concept under an even more general view. The probabilistic approach, the principle of superimposition of "states," of indeterminacy, can only prohibit any microscopic entity from having a defined trajectory. It is an "obvious" consequence that can no longer be denied. In the experiment of the two slits, this is perfectly represented. The single-photon (or electron or microscopic entity or — if we prefer —something comparable to Planck's size) travels all the possible trajectories between the source and the detector (always remember the QED and its language).

Going through all the possible trajectories, the photon also meets the two slits and, after passing through both, interferes with itself producing the light and dark bands of the interference (the QED explained it very well with the probability amplitudes). As Bohr used to say: "When the photon passes through slit A, this determines a certain possible world: world A; when it passes through B it builds world B. In the case of the double-slit, the basic concept is that both these worlds coexist, one superimposed on the other." By closing a slit we destroy an entire world. I know, very well, that it is almost impossible to believe that a single object can have the capacity to be in the SAME moment in two different places (simultaneity, poor Einstein...), but this is exactly what happens and the experiments prove it without a shadow of a doubt!

At this point, we fall back on the doubt that we have already formulated and which has been highlighted in the case of Schrödinger's poor cat. The objective reality of the microcosm (which is always what composes the Universe) depends or not on the "choices" made by who observes an experiment? The "simplest" vision, from a certain point of view, seems to demonstrate that if the observer decides to reveal the position of a particle, it ceases instantly to exist in its velocity dimension and vice versa. Everything is reduced, therefore, to the definition of observer and on this, we are still on the high seas, even if Feynman gives a rather ambiguous but sharable explanation: only the event changes. A way, however, to turn cunningly around the real underlying problem.

However, this can be addressed in several ways. First of all, the Copenhagen School. The role of the observer is something that cannot be disconnected from what is being observed. In other words, the observer and what is observed are part of the same system. The choices of the former determine the characteristics of the latter. This means that matter, before being measured, lives in a superimposed state. Only the intervention of the observer produces the so-called "reduction of the wave function" or collapse as one might say. Only in this way the superimposed state presents itself as a material entity with certain characteristics.

This paradigm, still completely open to both experimental and theoretical speculations (an experiment that is very difficult to perform since any experiment involves the intervention, sooner or later, of an observer) makes us better understand why Schrödinger, the "father" of the wave function, formulated the paradox. As we have repeatedly noted, Heisenberg's principle plays a fundamental role. It's the real central pillar of the whole construction.

It's worth mentioning its most shocking extrapolation for the entire Universe. It is not only true for speed (I should say "momentum" since mass is not as constant as it seems. Einstein's relativity has to coexist with QM, let's always remember that), but also for other, even more, decisive quantities: energy and time. In simple terms, this means that for very short times (and we always refer to Planck's magnitudes, in this case, his time), the law of energy conservation can be violated. Something really absurd (again), since we all know that in "our" world nothing is created and nothing is destroyed. How many times have we repeated it?!?

Once again, though, things, on a microscopic scale, don't just work like that. The state of indeterminacy that exists between energy and time can cause (indeed it certainly causes) energy fluctuations of a certain system in very short intervals. Intervals of billionths of trillionths of a second, in which an electron and its hostile positron friend can suddenly appear out of nowhere (the QED has shown us this clearly enough), unite and vanish. Even the most certain laws become indeterminate.

We don't think it's just theory since the lab experiments have confirmed it without a shadow of a doubt. Some might say, "Yes, all right, but they are fluctuations so small and short that they certainly cannot affect the reality of the Universe." And, instead, as it is easily predictable, they affect it a lot! It would be enough to think only of the tunnel effect (which we will deal with separately), capable of making a star live.

The void is like a huge ocean which, seen from a great height, seems to be perfectly calm and uniform. But if we'd go down on a small boat, we'd notice that it's furrowed by continuous and even violent waves. There, quantum particles can appear and disappear into thin air. We have returned to the fluctuations of the void, the sea of Dirac, the existence of positrons, asymmetries, the birth of the Universe, and inflation. The QM is a snake that bites its tail all the time. Yes, we can conclude that the very space that seems empty is the place where the most violent physical phenomena take place! Its very characteristics could force black holes to return to us the energy that was apparently swallowed and erased.

I'd say we can end these considerations here. They just tried to give a conceptual framework as general as possible and build "bridges" quite stable with what we have described in other articles and, in particular, in the QED exposition. I hope to have provided a little help in this regard, although we could talk and write for hours and hours and always revolve around the fundamental points and doubts.

What I ask of those who begin to ajar the door Alice has passed through is to communicate her feelings and emotions to the youngest. The future world is more and more in the hands of the QM, and it is right that young people form themselves with these concepts, in a similar way to what they do with those of force, speed, acceleration, angular momentum, etc. of classical physics and relativity.

If the school doesn't want (or can't) do it anymore, we try to do it ourselves. More importantly, let's make sure, in our small way, that young people can finally understand that they exist regardless of whether they have recorded themselves with their cellphones. We are part of nature, governed by the microcosm and the QM, we still exist. The cellphone gives us virtual reality, useful and extraordinary precisely because it will be increasingly linked to QM and — perhaps — to the rules and language still hidden in our brain. But it is an imperfect and partial simplistic representation of nature. It is not nature and neither is reality! It's just the execution of games without any mental effort. A way of representing it that can be useful to those who understand nature, but that can never be nature. In this context, believe me, the QM is much less absurd...

If we cease to exist, but we too become just a virtual representation, what sense would it make to understand what exists around us? But also, and above all, can we ever know it? It will probably cease to exist as well. At least for our shrunken brain leftover and unable to act autonomously. The Copenhagen school specifically says that we are one with matter and if the observer-actor is missing, the whole system collapses.

MQ not only as Science but as a school of life.